How to be an engineer for Smart kids.

kiddies simple steps to becoming vibrant and creative engineers.

Dylan Austin

Content

Introduction to Engineering
- Why was this book written? (A message to parents and guardians)
- What is Engineering?
- Different types of Engineering to make a choice.

The Marvelous World of Machines
- Simple Machines Explained
- How Machines Make Our Lives Easier

The Power of Electricity
- Lightning and Electricity: A Surprising Connection
- How Switches and Circuits Work

How to become an engineer for kids

Building Blocks of Structures
 - The Secret Behind Strong Bridges

Mysteries of Materials
 - Why Some Things Float and Others Sink
 - Amazing Properties of Different Materials
 - How to build a floating wonder.
 - List of Thoughtful and inventive ideas

Fun with Forces
 - Push and Pull: Understanding Forces
 - Gravity - The Force That Keeps Us Grounded

The Incredible World of Robotics
 - How Robots Listen, See, and Move

Engineering in Everyday Life
- Gadget Magic: How Phones and Tablets Work

Inventions and Innovation
- How Ideas Become Inventions
- Meet Famous Kid Inventors

How to become an engineer for kids

Introduction to Engineering

WHY WAS THIS BOOK WRITTEN?

(A message to parents and guardians)

Just like Pablo Picasso once said, *every child is an Artist*, I believe there's an innate trait of Engineering in every child. Here's why I believe the above statement. I remembered when I was a child, I used to play with self-automated toys such as mini cars and bikes. And to be frank with you, I wanted to tear them apart every single moment i played with them to see what they were made up of.

I was curiously driven to see the *thing* that was causing these toys' motion. Momma would say, 'Don't do that,' because if you do, you'll ruin your toy. But we all know how the story goes in every child's life; they'll make sure they come back to do the exact thing they were told not to do. Why? Because of curiosity.

I took apart those beautiful, well-engineered pieces of the toys and I found some components within body. One of them was the electronic DC motor. The electronic DC motor, whereby DC stands for Direct current, was the tiny giant that does the spinning, connected to some other components like loops causing some mechanical behaviors that propel the toys into movement. The feeling of finding this DC motor was heavenly, as one could see the expression on my face. It was like I'd won the jackpot.

However, this would be the last time the toy will ever function again. And here's the part where this book will inspire kids to think beyond just being curious into taking action on a unique idea to birth something new. What if I thought that the DC motor I removed could have been used to build a different kind of mechanical machine that helps us in the kitchen, at school or used to produce a fancy home appliance?

In the famous statement of Pablo Picasso above, he added that *the problem is how to remain an Artist once we grow up.* It is on this note a child's passion has to be cared for, guided and nurtured from the onset. So that instead of ending up in the curiosity stage, they will go beyond to establish something unique—a problem-solver solution.

In the great world of possibilities, understanding how things work opens up a lot of creative and smart ideas for kids. When kids explore how everyday things like toys or gadgets function, it's like going on a cool adventure. It not only helps them understand the world but also makes them curious and clever. That's why it's great for kids in this generation to learn engineering – it teaches them to figure things out, create stuff, and look at the world in a new way.

The essence of this book is to inspire kids to think beyond the usual ideas of the already running techs in town and spark their creative ability to super work on existing ideas or birth a new one using the creative thinking techniques listed in this book.

In this book, kids will find out that engineering isn't just about machines and electronics; it's a tool that helps them understand the interesting stuff around them. By knowing how things they use every day work, they learn to think carefully and solve problems, which is super helpful for what they want to do in the future.

Engineering is also like a special kind of creativity. It makes kids think beyond what's on the surface, play around with things, and try new stuff. This way of thinking helps them handle challenges and come up with new ideas, which is really useful, not just for engineering but for many other things too.

Parents, imagine how much your child's way of looking at the world could change. Picture them not just watching things happen but being a part of making things happen. Seeing the world like an engineer sparks a big interest in learning, and that helps them do really well in different areas.

Think of this book as a guide to help your child see a future where they don't just use technology but also make their own. Imagine watching your child confidently deal with a fast-changing world, being good at not just adjusting to it but also making a difference.

The message is pretty clear - get into engineering with your child, find out the cool secrets of everyday things together. Desire to make a place where questions are okay, being curious is good, and coming up with new ideas is awesome. Give your child the tools to not only understand the world but also do things that change it.

Buying this book isn't just a nice thing for your child; it's an investment in their future. An investment in their development. It's about giving them the skills and way of thinking they need to do well in a world where coming up with new ideas or improving on an existing one is really important. Come along on this exciting journey where learning is like an adventure, and every challenge is a chance to grow.

In the pages ahead, let your child's curiosity and the fun of finding things out guide them. Get them excited about engineering, and together, let's go on a journey of creativity, coming up with new ideas, and discovering amazing things. The time to start is now, and there are so many possibilities waiting. Enjoy the wonders of engineering with your child, and let's explore a world full of creativity, smart thinking, and endless discovery!

What is Engineering?

For kids, engineering seems like a creative superhero life. It's about using your imagination and creativity to create amazing things that can change the world around you. Imagine being the master behind creating cool gadgets, creating beautiful designs, and making everything better!

When we talk about engineering for kids, we are talking about being super inventors. Remember the days of playing with blocks and connecting objects to create amazing designs? Engineers do something similar, but on a larger scale. They use their imagination to design and build things like bridges, cars, and even robots! It's like turning your enjoyment of games into a powerful force for good deeds.

Imagination engineering is an amazing skill that allows you to turn your wildest ideas into reality. Have you ever dreamed of flying in the sky? Engineers build planes and rockets to make dreams come true. Have you ever wondered how your toy car moves? Engineers build powerful engines for them. So when you become an expert, you become an agent of possibilities, turning dreams into reality.

Now let's go a little deeper. Imagine you are on the playground and you see that the swing is broken.

Many kids might feel bad about this, but experts like to see it as an exciting challenge. "Can I fix this swing and make it swing higher?" They start to think. This is the nature of the correct problem in engineering as an interesting theory waiting to be solved.

Engineering is like a password that opens the door to the world of thought. It's not just about productivity; It's about understanding what products work and making them better. Think of yourself as an explorer looking for the unknown; Engineers do the same thing but use machines, buildings and technology. Being curious is about asking questions and finding smart solutions.

Now let's put on our engineering thinking helmets and explore the world of experts. Engineers are like the architects of the future. They design and build everything you see around you, from the house you live in to the playground where you play. It's like being the commander of an aircraft, teaching the class ahead and making sure everything goes well.

Just like your favorite superheroes have special powers, engineers also have special skills. They are skilled at using math and science to solve problems and create amazing things. It's not about being the fastest or strongest; It's about being the smartest and most creative person possible. Engineers are dreamers who make dreams come true.

So when you think of engineering, think of being a hero who makes the world better, cooler and more fun. This isn't just a job; It is a great power that you can use to shape the future. So put on your thinking hat, grab your creative box and get ready for a fun engineering adventure; Every problem is a challenge waiting for you Smart problem solver!

Types of Engineering

Hey, future engineer! Are you ready to enter the exciting world of engineering? Engineers use their brains and creativity to solve problems and create amazing things that help people. There are many types of Engineering to explore, and each one has a unique power waiting for you to discover!

Mechanical Engineering: Think of the possibilities of designing and building all kinds of amazing machines and devices. This is the job of mechanical engineers! They build things like robots, cars, planes, and even roller coasters. If you are the kid of kid that love taking things apart and put them back together to see how they work, this might be the best Engineering department for you.

Electrical Engineering: Have you ever wondered how electronic devices around you, such as phones and computers, work? Electrical engineers run these! They create the circuits and machines that give life to electricity, allowing us to communicate, play games and watch movies.

Civil Engineering: Dreaming of building skyscrapers, sturdy bridges and beautiful tunnels? Civil engineers achieve this! They design and build large structures like roads, dams, and even parks that make our cities work. If you're passionate about making the world safer and more beautiful, civil engineering may be for you.

Computer Engineering: Are you a computer genius and love coding and programming? Computer engineers are like geniuses! They design and build the hardware and software that make up our digital world, from smartphones and video games to artificial intelligence and cybersecurity.

Aerospace Engineering: Do you dream of soaring like a bird or exploring space like an astronaut? Aerospace engineers make these dreams come true! They design and build planes, helicopters, satellites and rockets that defy gravity and take us to new heights.

Environmental Engineering: Are you passionate about protecting the Earth and its beauty? Environmental engineers work to keep air, water and land clean and safe for everyone. They develop solutions to pollution, climate change and other environmental problems facing the world today.

Biomedical Engineering: Do you want to use science and technology to improve human health and save lives? Biomedical engineers, like medical researchers, design and manufacture medical devices, organ systems, and disease treatments. If you love biology and want to make a difference in healthcare, this may be your greatest strength.

Chemical Engineering: Want to know how chemicals are different and how they are converted into useful products? Chemical engineers use the power of chemistry to design processes and machines used to make products ranging from food and medicine to petroleum and plastics.

Materials Engineering: Are you interested in exploring different materials such as metals, plastics and ceramics? Engineers use materials to learn how materials behave and how they can be used to make everything from buildings and bridges to football equipment and smartphones stronger, lighter and more durable.

Industrial Engineering: Want things to be better and smoother? Engineers include business leaders who optimize processes and procedures to increase efficiency and reduce waste in factories, businesses, and organizations.

Remember, there is no right or wrong choice when it comes to choosing energy engineering. The most important thing is to follow your passions and interests. So go ahead and explore the exciting world of engineering! who knows? You may find your own superpowers along the way.

The Marvelous World of Machines

Simple Machines Explained

---

let's take a journey into the basic world of engineering machines in a way that sparks curiosity and creativity in the minds of young learners.

Imagine a world filled with incredible machines, like the ones you see at the playground or even in your toys. These machines, my friends, are like magical friends that help us do all sorts of things. Have you ever wondered how they work? Let's explore together!

Firstly, let's talk about levers. Now, levers are like the superheroes of movement. Think of a seesaw at the park. When you sit on one end, the other end goes up. This happens because levers have a special point called the fulcrum. When you push down on one side, it makes the other side go up. It's like magic! Engineers use levers to make lifting things

easier, just like when you play on a seesaw with your friends.

Now, picture a wheel and axle. This dynamic duo makes moving things a breeze. You know how your bike or toy car has wheels? When these wheels spin, they make it easy for you to go forward. It's like having a friend who helps you move without much effort. Engineers design cars and bikes using this fantastic idea, making them roll smoothly on the road.

What about gears? Gears are like special wheels that work together. Have you ever played with a toy robot and turned a handle to make it move? That's the magic of gears! When one gear turns, it makes another gear move. Engineers use gears to control how fast things go and how strong they are, just like when you play with different speeds on your toy cars.

Now, engines are like the hearts of machines. Imagine your favorite toy coming to life – that's what an engine does for many machines. In a car engine, fuel and air mix together, creating a powerful explosion. This explosion makes the car move! It's like a secret recipe for making things go zoom. Engineers are like chefs who cook up this special potion to bring machines to life.

Next up, pulleys. Pulleys are like little helpers that make lifting things a breeze. Picture a flag going up a flagpole. A pulley uses a wheel and a rope to lift things with less effort. It's like having a friend who helps you carry heavy bags up the stairs. Engineers use pulleys to lift heavy objects and make tasks easier for all of us.

Now, think about inclined planes. These are like ramps you see at the park. Have you ever rolled a toy car down a ramp? It goes so fast and smoothly! Engineers use inclined planes to move heavy things by making it easier to push or pull them. It's like having a slide for your toys to make them move effortlessly.

And then there's friction. Imagine sliding down a slide at the playground - it's so smooth! Now, think about sliding on sandpaper - it's a bit rough, right? Friction is like the slide's texture. Engineers use friction to control how things move. It's like choosing the right slide for your toys to make sure they go at the perfect speed.

But guess what? Sometimes, these machines go on an adventure called "Failure." It's not a bad thing! When something doesn't work, engineers become detectives. They investigate, learn, and make things better. It's like solving a puzzle to make sure the machines work perfectly. So, my little inventors, don't be afraid of failure. It's just a stepping stone to creating something even more awesome!

So, keep dreaming, keep playing, and who knows, maybe one day you'll be the superhero engineer who designs the coolest machines for everyone to enjoy. The world is full of possibilities, and you have the creativity to make it even more amazing!

How Machines Make Our Lives Easier

My young friend, machines are like consultants who make our lives easier and happier every day! Think of a world without them; No cool toys, no houses, and definitely no adventures. Let's look into the beautiful world of machines and discover how they've been working magic for years.

What are these machines, you ask? Machines are like smart children created by designers. Designers are like modern-day wizards who use their brains and creativity to create things that make life beautiful! These machines can be as simple as a bicycle or as complex as the high-tech equipment used by adults.

Now let's go back in time! In the past, people had to do everything on their own. Imagine having to do all the laundry without a washing machine or traveling long distances without a car. Seems like a lot of hard work, right? But then professional engineers came along and thought: "What if we built machines that could lift heavy loads for us?"

Let's move forward today! We now have awesome super smart machines everywhere. Look at your home; There is a washing machine and with the touch of a button your clothes are clean and fresh. Your freezer keeps your ice cold and your microwave keeps your snacks hot and toasty. These machines are like little helpers in your home, making everything easier and more fun!

Let's talk about creativity! Remember that engineers are finding new ways to make machines cooler. Think about your favorite movies; someone has to design them and then bring them to life using a machine! Who knows, maybe one day you will create a game that everyone will want to play!

Now let's think about the future! Think of machines that can take you to school in a flash or help you explore the depths of the ocean without getting wet. These are the types of mechanical engineers currently working! What happened? You could be the genius who creates the next great machine that will surprise everyone

Comparison is fun, right? Consider this: Doing homework without a computer is like trying to do hard work. Riding a square bike - not smooth, right? Machines makes everything faster, easier and more exciting!

Think about what you like to play! Do you remember role playing with your friends? Engineers are like the coolest playmates ever. They create machines that can make your wildest dreams come true. Next time you're building a castle or creating a hidden treasure, think about how technology and engineering can help make those adventures even more exciting!

Anyway, little adventurer! The machines and the bright hearts behind them are always working to make our world a better place. From simple gadgets to high-tech marvels, these magical aids make our lives easier and more feasible. So next time you see a machine, think about how it was designed as a piece of engineering magic that will make your life a little better!

Imagine, explore and create because the world is waiting for you to create incredible things!

The Power of Electricity

Lightning and Electricity: A Surprising Connection

There is an incredible connection between lightning and electricity that we may barely see with the naked eye, and it all started with the curiosity of some smart people throughout history. Long ago, scientists, like today's explorers, began a journey to understand the mysteries of nature.

One of these pioneers is the American scientist and inventor Benjamin Franklin. In the 18th century, he conducted an experiment with a kite during a storm. Imagine flying a kite in the rain, not for fun, but to uncover the secret of lightning. Thanks to this experiment, Franklin proved that lightning was a form of electricity and paved the way for future discoveries. Imagine a world without lights, computers or smartphones; everything uses electricity. Thomas Edison made the world brighter by inventing the light bulb in 1879. Kids, think of Edison as a role model who brings light to darkness!

As people delved deeper into the electric field, engineers like Nikola Tesla emerged. Just like it could be you renovating or working on an existing idea to perfect it, Tesla's products and partnerships have changed the way we use electricity. Consider that he is now a wizard, weaving magic to power our homes and devices. His work laid the foundation for alternating current (AC), which we use in our daily lives.

Now let's connect the dots with the junior engineers who do this job. Electricity is like an invisible pillars that supports our world. It flows through the wires like a mighty wave of the river, bringing our devices to life. Inventors, like today's geniuses, work tirelessly to ensure the safe and effective use of this magnificent power.

Think of engineers, the designers of the electrical world. They build the systems that make up our homes, schools, and parks. It's like creating a big puzzle where all the pieces have to fit together for the magic to flow. Guys, while you are preparing the blocks to build the towers, the engineers are preparing the cables and materials that will support our city.

Now let's add some creativity to the mix. Imagine being able to control electricity using your imagination! Close your eyes and imagine a world where you create tools powered by the power of your mind. That seem impossible right? But we've seen the possible emerging from the one time pronounced impossible. Isn't this great? Engineers are like dreamers, they turn those ideas into reality by the time they assess the idea with questions such as how, why, what, etc.

When we wonder about the concept of electricity, let's not forget its important role in improving our lives. From electrical equipment used in school to lighting your favorite playground, electricity is a quiet man. He is like a friendly friend; always ready to brighten your day and make your experience more enjoyable.

But as in every story, there are difficulties to be overcome. Kids, think about what you've done to find a better way to generate electricity without harming the environment. This is where your creativity comes into play. What if you could make a device that captured the energy of the sun or wind?

In short, the story of lightning and electricity; It is the story of curiosity, discovery and innovation. This is a story that connects the past, present and future. So little engineers, as you play with your toys and explore the world around you, remember that you have the power to create the future. Be inspired by electricity, dream big

How Switches and Circuits Work

What is a switch?

A switch is a tiny conductor that can open or close the pathway for electricity. It's like a door that, when closed, lets the energy flow, and when open, interrupts the flow. This simple yet ingenious device is the key to controlling the power that surges through our gadgets.

In your everyday play, think about the light switches in your home. When you flip the switch up, the circuit is complete, and electricity flows, lighting up the room. Flip it down, and the circuit breaks, leaving you in the cozy embrace of darkness. It's like having a superpower at your fingertips - the ability to command the invisible force of electricity.

Now, let's unravel the mystery of circuits. A circuit is a looped pathway that allows electricity to travel from its source, through various components like switches and lights, and back again. Imagine it as a journey that electrons embark upon, bringing energy to all the devices that make your world sparkle.

Step by step, let's explore how switches and circuits work their enchantment:

Energy Source: Every circuit begins with a source of energy, like a battery or an outlet. This is where the magic begins.

Pathway for Electrons: The electricity travels along a pathway, like a winding road for electrons to zoom through. This pathway is usually made of conductive materials that allow the flow of electricity.

Switching it On: When you flick the switch on, it closes the circuit. Imagine it as the moment the drawbridge lowers, allowing the electrons to cross and bring life to the devices connected to the circuit.

Device Activation: As the electrons flow through the circuit, they power up the devices connected to it - whether it's a light bulb, a fan, or a toy. The magic of electricity transforms into tangible actions and delightful experiences.

Switching it Off: Now, imagine the switch as a spellbook. When you cast the spell to turn it off, the switch opens the circuit. The drawbridge rises, blocking the pathway for electrons, and the devices rest in a state of slumber.

Completing the Circle: For the magic to continue, the electrons need a complete circle. They return to the energy source, completing their journey and preparing for the next adventure.

In your playful exploration, consider creating your own circuits with simple materials like paperclips, wires, and batteries. Build bridges for electrons to dance across and discover the joy of controlling the flow of energy with your homemade switches.

As budding engineers, you have the power to design circuits that not only bring joy to your play but also contribute to the marvels of our technological world. Think about ways to make circuits more efficient, using renewable energy sources or creating smart circuits that respond to your commands.

So, my young wizards of electricity, as you flip switches and weave circuits in your imaginative play, remember that you hold the magic wand to illuminate the world around you. Embrace the wonders of switches and circuits, and let your inventive spirit light up the path to a future where your creations shine brightly with the magic of engineering.

Building Blocks of Structures

The Secret Behind Strong Bridges

Building strong bridges is like creating massive puzzles that connect cities and communities. Imagine a giant game where engineers use special materials (called materials) to create these incredible structures. Let's look into the secrets behind the magic of building bridges that can withstand tons of weight and last for years.

Secret 1

Choose the Right Materials

When engineers begin the adventure of building a bridge, they choose their materials like a superhero chooses glasses. Metal and stone are the most preferred due to their incredible strength and durability. These materials together form a composite structure that allows the bridge to support heavy loads.

Secret 2

Understanding the Force

Meet powerful enemies such as gravity, wind and traffic. Engineers think about and understand these forces in order to build bridges that can withstand them. Just like a house construction need a strong foundation, bridges need a strong foundation to withstand the constant force of gravity.

Secret 3

Shape Matters

Have you ever noticed that bridges can come in all shapes? Like artists, engineers carefully select shapes that distribute weight evenly. The arched bridge is like a bow on a string, strong and supportive. The bridge beam acts like a scale, distributing the weight along its length. Each picture is a puzzle piece, making the bridge strong and durable.

Secret 4

Limitations and Challenges

Architects need to consider how much weight the bridge can support, and sometimes "No, it's too heavy!" they have to say. Storms and earthquakes can be formidable opponents, but architects build bridges with these challenges in mind to withstand them. a powerful force of great Nature.

Secret 5

Different Types of Connections

Bridges have special abilities depending on their species. Suspension bridges are like giant spider webs stretching over a long period of time. Truss bridges use a triangular mesh to represent the load. The bridge is like a harp measuring its weight in the sky with beautiful lines. Each species has their own superpowers that lead to the diverse world of Bridges.

Secret 6

Build Your Own Bridge

This is where creativity soars! Kids can unleash their inner engineers by building miniature bridges using everyday materials like stickers, tape and glue. It's like doing your own puzzle. By trying and learning, you will become a mini expert creating a path for small cars or works of art. Who knows, you may support a future professional!

Secret 7

Compare to Everyday

Think of the bridge as a large dining table. The legs act as a foundation that supports the weight of the table, just like the foundation of a bridge supports the weight of the table. The desktop is like a bridge deck where everything happens. Just like you can't put too much heavy metal in one place, engineers made sure the weight of the bridge was evenly distributed.

In short, the bridge is a great adventure combining science, art and creativity. Engineers are like modern-day geniuses who use their knowledge to connect the world and make our lives better. So next time you cross a bridge, remember the superhero-like effort it took to create that art, and who knows, maybe one day you'll create the crossing bridge yourself. Become a legend in your community!

Mysteries of Materials

Why Some Things Float and Others Sink

Imagine a small boat sailing gracefully on the water or a rubber duck floating effortlessly in the bathtub. What makes these objects stay afloat, defying the pull of gravity? To understand this phenomenon, we must delve into the very fabric of matter.

At the heart of the matter lies the concept of density. Density is a measure of how tightly packed the particles in a substance are. Think of it as a crowd in a playground - if the children stand close together, the crowd is dense, but if they spread out, the crowd becomes less dense.

Objects float or sink based on their density compared to the density of the fluid they are placed in, often water. If an object is less dense than the liquid, it floats; if it's denser, it sinks. Take note that my dear young Engineer. Because this is the basis principle at which your invention will work out perfectly. Picture a rubber duck - its hollow body makes it less dense than water, allowing it to bob on the surface with ease. On the other hand, a heavy rock, being denser than water, obediently sinks to the bottom.

Now, how can we engineer objects to float or sink? It's all about manipulating density. For things to float, we can create spaces within objects, making them less dense. A boat, for example, a boat has a hull filled with air pockets that decrease its overall density, enabling it to stay afloat. These air pockets also serve as compartment. If you have seen the movie Titanic youd know that you can actual boat any boat with such aestheric compartments and still flow perfectly on a high sensed water body. I encourage the young minds to imagine constructing a boat out of lightweight materials with cleverly designed compartments to trap air, giving it the magical ability to float.

Consider the concept of submarines too, fascinating vessels that can float or dive underwater. Through the clever use of ballast tanks, engineers control the submarine's density, making it rise to the surface or descend into the depths. It's like having a secret trick to control whether your toy can swim on the water's surface or dive into the depths of the imaginary sea.

But why should we care about these floating and sinking wonders? Because this understanding unlocks the door to endless possibilities for improving lives! Imagine a world where floating structures support entire communities in flood-prone areas, providing safety and shelter during challenging times. Picture bridges constructed with materials that adapt to water levels, ensuring resilient transportation.

As young Engineers, think about designing devices that can detect changes in density, helping identify water pollution or ensuring proper mixing of ingredients in cooking. What if our everyday objects could transform based on the environment, just like a boat adjusting to the water's surface?

Engineering, the art of making things work, empowers us to turn these imaginative ideas into reality. It's not about superpowers but about harnessing the forces of nature. Through engineering, we can create solutions that make a positive impact on our lives and the world around us.

So, dear young minds, as you play and explore, remember that the mysteries of materials hold the keys to innovation. Embrace the wonders of density, engineer with creativity, and dream of a future where your inventions contribute to the betterment of lives. The journey of discovery begins with a simple question: why do some things float and others sink? Unlock the secrets, and let your imaginations soar!

Amazing Properties of Different Materials

There is a fascinating array of materials in the world around us, each with unique properties that add thoughtfulness to our daily lives. One of the most interesting phenomena is the ability of some objects to float. Let's begin a journey to uncover the surprising properties of different products, discover the reasons behind their airiness, and discover how these products can inspire new ideas to improve our lives, especially in the engineering industry.

Understanding buoyancy:

At the heart of the problem is a force called buoyancy, which causes objects to float or sink in fluids such as water or wind. Imagine yourself in a swimming pool with a beach ball in your hand. The ball floats effortlessly in the water, showing the effect of buoyancy. This phenomenon is governed by Archimedes' principle, which states that an object placed in a fluid produces an upward force equal to the weight of the fluid it displaces.

Materials and their buoyancy properties:

Woods
Why it floats: The density of wood is less than water, which allows it to absorb enough water to support its weight.

Everyday Analogy: Imagine a wooden toy boat sailing smoothly on a lake; This is like a masterpiece of natural engineering.

Plastics:
Why They Float: Many plastics are less dense than water, which allows them to float.

Everyday Analogy: Consider a plastic bottle bobbing in the bathtub, demonstrating how intelligence can make everyday objects useful and powerful.

Cork:
Why it floats: Cork is full of air pockets, making it less dense than water and allowing it to float effortlessly.

Everyday Analogy: Think of a mushroom floating in a glass of water; like a little craftsman working wonders in the forest.

Miracles of Engineering: Buoyancy in Design

Engineering Insights: Engineers use the principle of buoyancy to design boats that can carry heavy loads across bodies of water.

Everyday example: Consider a ship carrying cargo across the ocean, using the principle of transportation to achieve a good result.

Innovative Structures

Engineering Insights: By understanding buoyancy,

engineers can design structures such as pontoons that can cross bodies of water. Water on bridges connecting two beaches for seamless travel.

Thinking Beyond
Inventions of the Future:

Inspiration for Innovation: Understanding why objects float opens the door to innovative technologies such as floating cities or floating settlements. Door.

Everyday Analogy: Imagine a future where communities live on islands and use innovations to change the environment.

Environmental Solutions

Designed for a Better Planet: Inspired by buoyancy, engineers can create solutions such as floating waste products to solve water pollution.

Everyday Analogy: Consider a floating device that effortlessly collects waste, demonstrating that engineering can solve environmental problems.

In summary, the incredible properties of floating objects are a testament to the balance of energy and density in our world. By understanding these principles, young people can imagine a future where work is a great way to enrich life.

Encourage children to explore the concept of buoyancy through projects such as building their own pontoon boat using everyday materials. In this way, they not only know the principles of science, but also develop the curiosity that will inspire them to become professionals in the future, creating inventions that will improve our lives and protect our planet. Let's raise a generation that not only values everyday issues, but also dreams of new solutions for the future.

How to build a floating wonder.

Steps to Build a Floating Wonder

Material Selection: Choose materials that are light and can easily displace water, like foam or hollow plastic.

Design: Plan a shape that maximizes the volume without adding unnecessary weight. A wide and flat design is often effective.

Testing: Experiment with different designs and materials. Test your creation in a small pool to observe how it floats.

Functions Beyond Fun:
Now, let's think beyond the joy of floating paper boats in puddles. Materials with buoyancy serve crucial purposes in our everyday lives.

Life Jackets: Imagine if our life-saving jackets were heavy and sank. Engineers design them with buoyant materials to keep us afloat in emergencies, ensuring safety on water.

Boats and Ships: The vessels that transport goods across oceans are carefully engineered to stay afloat. Buoyancy is a vital factor in their design, allowing them to carry heavy loads without sinking.

Swimming Aids: Pool noodles and inflatable floaties are crafted with buoyant materials, making learning to swim both safe and enjoyable.

List of Thoughtful and inventive ideas

Young minds are encouraged to think about how the concept of floating can inspire inventions that improve lives.

Floating Farms: What if we could create floating farms to grow crops in areas with limited land? Buoyant structures could revolutionize agriculture.

Floating Cities: Imagine entire cities designed to float, providing sustainable living solutions in regions prone to flooding. Engineers could build resilient communities that adapt to changing environments.

Floating Clean-Up Bots: Buoyant devices could be designed to navigate water bodies, collecting plastic waste and keeping our oceans clean.

Everyday Play to Engineering Marvels

From playing with floating toys in the bath to envisioning future inventions, the journey from childhood curiosity to engineering marvels is interconnected. Children can grasp the concept of buoyancy through hands-on activities, sparking their interest in the world of materials and engineering.

In essence, the mystery of why some things float and others sink unravels through the ingenious language of materials and engineering. By exploring this enchanting world, young minds can unlock their creativity and begin to dream of inventions that promise a better, buoyant future for all.

Fun with Forces

Push and Pull: Understanding Forces

Gravity is the force that keeps us on the ground and is the fundamental concept that helps us understand the world. It plays an important role in force science, especially in terms of pushing and pulling forces. Let's understand the complexity of gravity and its implications for young engineering enthusiasts.

Understanding Gravity

Gravity is the force that pulls objects towards each other. Imagine dropping a ball; The ball falls to the ground due to gravity. This energy makes everything in the world. Sir Isaac Newton was a brilliant scientist who developed the law of universal gravitation, which explains how objects attract each other based on their mass. It is like an invisible bond that connects everything, from the smallest rock to the largest mountain.

How does gravity affect us

Gravity doesn't just affect objects, it affects us! That's why we stay on the ground and don't swim. This energy is universal and affects everything with its magnitude. Even the moon orbits around the earth due to the influence of gravity. So gravity is not just a force; It is the cosmic glue that holds the world together.

Effects of Push and Pull

Now let's connect the concepts of gravity and push and pull. When you jump, you use upward force (while) against gravity. But gravity pulls you back, providing a perfect example of the interaction between push and pull. Designers use these patterns to create structures and materials that can withstand these forces or use them to our advantage.

Gravity in Everyday Engineering

Consider a bridge—Engineers must understand gravity to ensure the bridge can withstand the load and absorb the force from cars and other loads. Similarly, a car moving forward encounters resistance (pull) and relies on the engine's power (push) to overcome it. All modern examples involve balance of forces, making gravity an important engineering principle.

Impact on Young Engineers

For aspiring young engineers, understanding gravity opens the door to incredible possibilities. Consider building tall, strong buildings that can defy gravity, or building cars that can carry cars and pull power onto the road. Like being a scientist, keeping things calm.

Think like a young engineer

Imagine this; You are building a tower from building blocks. Each block represents a force; Some pushing, some pulling. The tower must be solid, just like a real building. By trying different plans you will already feel like an engineer! It's about finding balance, just like engineers do when designing and manufacturing products.

Better Design, Better Living

Now let's connect gravity to the big picture. Imagine a world where engineers use their understanding of gravity to create products that make our lives better. It could be a safer car, a more sustainable building, or even a new way to produce clean energy. The possibilities are endless, and it all starts with knowing the force that keeps us on the ground: gravity.

In the exciting world of forces, gravity is a force that shapes our world and affects everything around us. Young engineers can use this power to improve our lives. By understanding the push-pull force, they can create a future where products not only resist gravity but also improve our lives.

The Incredible World of Robotics

How Robots Listen, See, and Move

Let's embark on a journey to uncover these magnificent creatures of men that hear, see and move, and learn about their great impact on our people.

Robots are equipped with their own version of these senses, just as you have eyes to see and ears to hear. Sensors act as their eyes and ears, allowing them to know the world around them. These sensors work together to collect data and understand their environment, just like the teams in your favorite sport.

Acoustic Engineering

Now let's see how robots listen. This is quit simple, just like you hear sound with your ears, robots use something called a microphone that acts like their super-sensitive ears. Intriguing right?

These microphones pick up sounds just like your ears do, they are designed to listen for specific frequencies, just like your favorite radio. Professional Engineers around the world are working hard to translate these "listening" abilities into robots so they can better understand and respond to different sounds around them. And this is already happening in some areas around the world today.

Visual Perception

Speaking about vision, robots have cameras they use as eyes. These cameras capture images and videos that allow the robot to "see" the world. Just like you recognize a friend's face, a robot can undergo training to recognize objects and people. It seems to give them special eyes to navigate the world and work with reality.

Their motion

Now let's talk about how the robot moves. Think of it like a painting or a toy car; Robots have their own "fixes" and "motors" just like their muscles. Engineers use mechanical engineering knowledge to create these components so that robots can move easily and accurately. It's like playing with building blocks; all the pieces come together perfectly to create a functional and functional experience.

Improving Life
Through being able to hear, see and move, we create tools that can actually improve our people Imagine a world where robots assist doctors in surgery, help save people in disaster areas, and even help explore space. It's not about having superheroes, it's about having a reliable partner who makes our lives better and safer.

The Age of Robots

Now kids, let's think about the future. As you grow, you will have the opportunity to be a part of the robot age. You can be an expert and inventor who creates the best robots. Just like playing with blocks or solving puzzles, you can use your imagination to come up with new ways for your robot to hear, see and move. This can lead to incredible productivity that we can't even imagine right now!

Learning by Doing

Think of the robot journey as a giant playground of ideas. Just like playing with toys, engineers try different things together to make robots better. You can think of it as an endless game where each attempt brings us closer to finding new ways to improve our lives and solve problems.

In short, kids, the world of robots is like a big adventure. Artists and inventors work tirelessly to create magical creatures that can hear, see, and move. As you explore your interests and hobbies, remember that you could be the future inventor revolutionizing robotics. So embrace the excitement, think creatively, and imagine the endless possibilities the robotic age will bring to improve our world.

Engineering in Everyday Life

How Phones and Tablets Work

Mobile phones and tablets have become an integral part of our daily lives, connecting all segments of society. The evolution of these devices is not only changing communication, but also the way we work, learn and play.

Mobile phone and tablet production includes interconnected devices in many countries. Manufacturers produce components such as processors, memory and displays from different suppliers. The components are then assembled with high-tech equipment and the final product is rigorously tested before reaching the customer. This complex process requires precision and coordination among different stakeholders in the global technology industry.

Ꭲhe development of mobile phones and tablets dates back to Alexander Graham Bell's invention of the telephone in 1876. Bell's innovation changes the way people communicate by allowing sound to be sent over long distances. The telephone has become a symbol of connection, allowing instantaneous conversations across geographic boundaries.

Fast forward to the end of the century, the communication

environment witnessed another change with the advent of mobile phones. Inventors and engineers envision a world where people can communicate wirelessly without the limitations of traditional phones. The first mobile phone, the Motorola DynaTAC 8000x, was released in 1983 and marked the beginning of a revolution.

The inspiration behind the mobile phone came from the desire for better accessibility and mobility. Innovators are trying to create tools that allow people to communicate on the go, challenging the idea that communication should be limited to certain locations. As technology has advanced, mobile phones have evolved from bulky devices with limited functionality to stylish, versatile smartphones with multiple functions.

The concept of easy-to-use technology is now attracting attention. Steve Jobs and his team at Apple played a key role in bringing the tablet to the forefront of consumer technology. Introduced in 2010, the iPad popularized the idea of a touch-screen tablet that could be used as a multi-purpose tool for business, entertainment, and education.

    The inspiration behind the tablet was placed in the vision of a device that could bridge the gap between smartphones and traditional computers. The tablet has many easy-to-use and easy-to-use features that appeal to a wide range of users, from professionals to students and consumers.

Before the widespread use of mobile phones and tablets, life was seen through different communication and business practices. Letters, telephone calls, and telephones are important forms of communication. The process of sharing information and doing business often requires an engine or a longer turnaround time.

The introduction of mobile phones and tablets has changed the way we communicate. Instant messaging, video calling, and social media have become an integral part of our daily interactions, instantly connecting people around the world. The ability to access information instantly has revolutionized the way we learn by putting information at our fingertips.

Mobile phones and tablets have revolutionized productivity in the workplace. Email communication has become very fast and mobile devices have made remote working easier. Professionals can collaborate on projects, access information, and participate in virtual meetings from anywhere; This changes the traditional way of working in the office.

Entertainment has also changed dramatically. Streaming services, mobile games and social media provide constant entertainment. The convenience of owning a portable device with a variety of entertainment options has redefined the way we consume media, making it more personal and accessible.

The era after phones and tablets has seen not only the benefits of digitalization but also the challenges of connecting. Concerns about screen time and its impact on mental health, particularly in children, have been highlighted. Balancing the benefits of technology with the need for face-to-face interaction and outdoor activities has become a priority for individuals and families.

Privacy issues are becoming more evident as private information is expanded and stored on these devices. Constant connectivity also creates problems in managing work-life balance; because the lines between professional and personal life are blurring due to the ability to work remotely at any time.

But the overall impact of mobile phones and tablets on people has changed. These tools streamline access to information, allowing people of all backgrounds to connect, learn, and participate in the digital age. Communication has improved and become more efficient, enabling faster response to international events and crises.

As a result, mobile phones and tablets have transformed from visual tools into indispensable parts of life today. Inspired by the desire to improve communication and mobility, complex manufacturing processes have given rise to devices that shape the way we work, learn and play. While acknowledging the challenges brought by the digital age, it cannot be denied that mobile phones and tablets in society have changed the way we connect and interact with the world around us.

Inventions and Innovation

How Ideas Become Inventions

Turning ideas into inventions is an exciting process that begins with curiosity and inspiration. Like everyone else, children can begin this creative process by observing the world around them and letting their imaginations run wild.

The first step in this happy journey is to pay attention to the world. By observing objects and events today, children can identify problems or areas that arouse their curiosity. This could be as simple as wondering why certain tasks are taking so long or why something isn't working properly.

Most ideas come from the original. Children should be encouraged to draw inspiration from everyday experiences. Whether it's a challenge they're facing or a desire to make things even more exciting, these little moments can spark ideas. For example, seeing the difficulty of untangling headphone cables can lead to ideas for wireless solutions.

Imagination is a powerful tool for children. Encouraging them to play and explore their imagination will help develop creativity. By imagining different situations or creating new uses of existing objects during play, children train their minds to think clearly and lay the foundation for positive thinking.

Teaching children different types of ideas can be an important skill. Simple things like mind mapping, connecting ideas to text, or asking "what if" questions can encourage positive thinking. These strategies help break problems into smaller pieces, making it easier for children to find new solutions.

Children need to understand that ideas can have a positive impact on the world. Children can understand the transformative power of ideas by sharing stories about the development of life, such as the invention of the wheel or the discovery of electricity. This experience provides a sense of purpose and motivation to think beyond self-interest.

It is very important to teach children that ideas are just the beginning. Thinking with an idea involves thinking of ideas that will make them a reality. For example, if a child needs a device to help the elderly, he/she must consider its material, design and functionality. This thought process helps bridge the gap between theory and real-world application.

Teaching engineering concepts in simple terms allows children to connect with problem-solving strategy. Engineering is about finding solutions to problems, just like problems encountered in daily life. Children can understand the practicality of these ideas by breaking down building blocks into examples, such as building a strong tower from building blocks or building a simple towable machine.

Explain the idea that development is designed to improve life and that responsibility can be improved. Whether it's easier homework or solving a traditional problem, children need to understand that their ideas can contribute to the well-being of others. This perspective creates a sense of purpose and understanding, encouraging them to think beyond personal gain.

In short, the language of turning ideas into inventions is a fascinating process that emerges with observation, inspiration, imagination and creativity. By teaching these concepts to children in a timely and relevant way, we empower them to become the innovators of tomorrow and create solutions that impact lives.

Meet Famous Kid Inventors

- Louis Braille (1809-1852):
  Louis Braille, who was blind at a young age, invented the Braille system when he was only 15 years old. His incredible writings opened the world of reading to the visually impaired, proving that determination and innovation have no age.
- George Neeson (1914-2010):
  Inspired by circus acrobats, George Neeson invented the trampoline in 1930, when he was

16 years old. Little did he know that his invention would become a sport in gymnastics. Joy for many generations around the world.

- Chester Greenwood (1858-1937): In 1873, 15-year-old Chester Greenwood invented earmuffs to keep his ears warm on cold days. Its simple yet practical construction has provided warmth and comfort to many ears this winter ever since.
- Philo Farnsworth (1906-1971)): Philo Farnsworth, often referred to as the "Father of Television", came up with the idea of electric television when he was 14 years old. His inventions revolutionized

entertainment and communication. , shapes the way we connect with the world through the magic of moving images.
- Anne Marie Imafidon (1990-present): >Anne Marie Imafidon stands out in technology. He is a math prodigy who passed his computer exam at the age of 11. As the co-founder of STEMettes, she works to encourage young women to pursue careers in science, technology, engineering and mathematics (STEM).
- Samuel Thomas von Sömmering (1755-1830):
German inventor Samuel Thomas von Sömmering Sömmering

invented the telephone when he was 12 years old. His curiosity for young people influenced the development of technology and laid the foundation for the future of communication.

- Gitanjali Rao (2005-present):
Gitanjali Rao, Time Magazine's 2020 Child of the Year, is a young inventor and scientist. At the age of 12, he built an instrument called Tethys, designed to detect organisms in water. Gitanjali embodies the power of youth to solve real-world problems.
- Alexander Graham Bell (1847-1922):
Before becoming famous for the invention of the telephone,

Alexander Graham Bell showed his creative spirit early. age. As a child, he established a communication with his deaf mother, which foreshadowed his later contribution to the world of communication.

Stories of famous child inventors show that age is no barrier to innovation. Their journey begins with curiosity, passion, and a desire to make a positive impact on the world. So kids, let these stories inspire you to dream big, explore your imagination, and maybe one day you'll be the young inventor who makes history.

The End

www.ingramcontent.com/pod-product-compliance
Lightning Source LLC
Chambersburg PA
CBHW070153230526
45471CB00002B/639